YOUR KNOWLEDGE HAS VALUE

rafaqat ali

Impact of vegetal cover on local water resources. A case study of Kanshi sub-watershed (Gujar Khan), Punjab, Pakistan

GRIN Verlag

Bibliografische Information der Deutschen Nationalbibliothek:

Die Deutsche Bibliothek verzeichnet diese Publikation in der Deutschen National-
bibliografie; detaillierte bibliografische Daten sind im Internet über http://dnb.d-
nb.de/ abrufbar.

Imprint:

Copyright © 2014 GRIN Verlag GmbH
Druck und Bindung: Books on Demand GmbH, Norderstedt Germany
ISBN: 978-3-656-69364-2

This book at GRIN:

http://www.grin.com/en/e-book/276214/impact-of-vegetal-cover-on-local-water-
resources-a-case-study-of-kanshi

GRIN - Your knowledge has value

Der GRIN Verlag publiziert seit 1998 wissenschaftliche Arbeiten von Studenten, Hochschullehrern und anderen Akademikern als eBook und gedrucktes Buch. Die Verlagswebsite www.grin.com ist die ideale Plattform zur Veröffentlichung von Hausarbeiten, Abschlussarbeiten, wissenschaftlichen Aufsätzen, Dissertationen und Fachbüchern.

Visit us on the internet:

http://www.grin.com/

http://www.facebook.com/grincom

http://www.twitter.com/grin_com

Impact of vegetal cover on local water resources:

A case study of Kanshi sub-watershed (Gujar Khan), Punjab, Pakistan.

Ali, R & Anwar M.M.

Department of Geosciences and Geography

University of Gujrat

Corresponding author email: raafi_pu@yahoo.com

Ali, R(Rafaqat Ali) is the M. Phil student at the university of Gujrat. His initial work on the impact of agricultural practices in Mangla watershed. Anwar M. M (Muhammad Mushahid Anwar) is chairman of Department of Geosciences and Geography in University of Gujrat. He is an urban ecologist having his doctorate degree from Europe

Impact of vegetal cover on local water resources: A case study of Kanshi sub-watershed (Gujar Khan), Punjab, Pakistan.

Abstract

This research was conducted focusing the importance of vegetation in Mangla watershed for the assessment of its impacts on water resources. Satellite images of study area were analyzed using Arc GIS 9.2 and Erdas Imagine 9.1 for vegetation change detection and identification of local streams feeding Kanshi River. River discharge data was obtained from Surface Water hydrology department Lahore. Extensive agriculture practices, population growth, settlement patterns and brick industry have significantly affected the vegetation cover. Deforestation occurred in the past for agriculture and timber needs has changed the vegetation condition and hence rainfall patterns. Uncertain or extreme rainfall events and temporary drought condition are common. The discharge of Kanshi River is decreased by 44.15% during last 20 years with a significant decrease in water table of Gujar Khan City. Further development in watershed area, improper agriculture practices, unplanned settlements and deforestation should be completely banned to stabilize the ecosystem. Water harvesting structures should be constructed to fulfill the needs of local community for agriculture and domestic use.

Keywords: Gujar Khan, vegetation, watershed, river discharge, ArcGIS

1 Introduction

The exploitation of land and water resources to sustain an ever-increasing population inevitably involves the utilization for both urban and agricultural development of rural areas and the natural landscape. This process can result in profound changes to the flow regime of river basins that are so affected, the scope and magnitude of which have been investigated by means of experimental catchment studies (Sahin and Hall 1996). Rivers, and their vegetation, integrate the effect of all the environmental impacts acting on them, impacts both natural (e.g. rock type) and human (e.g. pollution, recreation). Human impacts vary in both their pressure, and in how that pressure is handled (Haslam 1995). Less precipitation has been an important factor driving the decrease in runoff and sediment discharge during 1999–2007. However, restoration and improvement of the vegetation cover may also have played a significant role in accelerating the decrease in annual runoff and sediment discharge by enhancing evapotranspiration and alleviating soil erosion (Xin and Yu 2009). Increasing agricultural practices and deforestation is resulting into the uncertain rainfall and changed pattern of river flows worldwide. Loss of vegetation cover not only contributes to less average rainfall and minimum delay time of arrival of run-off into local streams resulting in lowering of mean water table but it also accelerates natural hazards like erosion and flash floods in highlands especially. This study was conducted to identify the vegetation cover change and its impacts on local water resources.

1.2 Study Objectives

Main objectives of the study were to

> ➤ Identify the role of vegetation cover on local water resources
> ➤ Map the local streams contributing to Kanshi River.

1.3 Study Area

The study area consists of tehsil Gujar Khan of Rawalpindi district which is included in potohar region of Pakistan. The coordinates of the study area are 33°20'57.91" N, 73°19'33.06" E, 33°20'18.16" N, 73°27'18.63" E, 33°14'39.04" N, 73°17'53.36" E and 33°12'24.37" N, 73°25'58.11" E and Kanshi watershed is sub-watershed of Mangla watershed which is situated in tehsil Gujar Khan and tehsil Kahuta of Rawalpindi district. "Har", "Kurri", "Missa", "Guliana" "Phahna" and "Gulin" are sub-streams of Kanshi which after joining Kanshi drain into the Jhelum River at 33°14'54.59"N Latitude and 73°36'24.21"E longitude.

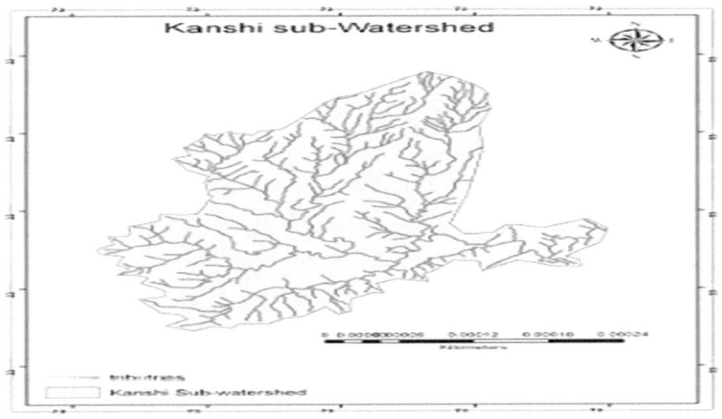

Map 1: Kanshi Sub-watershed and its tributaries
Source: Prepared on Arc GIS 9.2 from the topographic map of Rawalpindi

3

2. Materials and Methods

Satellites images (of Landsat TM 30x30 meter resolution) of 1992 and 2000 were obtained from World Wide Fund for Nature (WWF), Lahore, Pakistan for change detection in vegetation cover and analyzed using Erdas imagine 9.1 and Arc GIS 9.2.Satellite image of 10 x10 meter resolutions were obtained from a private organization "Jerse" to trace the local streams contributing to Kanshi river. River discharge of Kanshi was obtained to draw the discharge graphs from 1990 to 2010. A topographic sheet of Rawalpindi was used to delineate Kanshi watershed. Field visit was arranged to gain the first hand knowledge of present and past water table statistics, a questionnaire was prepared by keeping in mind the possible issues which came out after literature review and data acquisition. Questionnaire and interview data were used to identify the socio-economic condition of locals.

3. Data Analysis and Findings

Graph 1. Average water table from 1990-2010   Graph 2. River discharge from 1990-2010

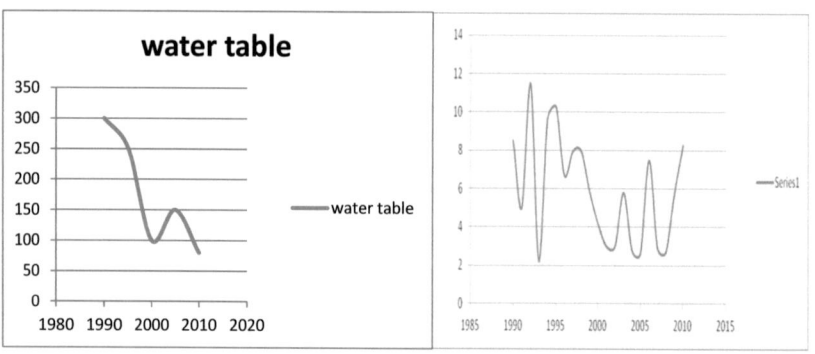

4

Map 2: Vegetation cover of 1992
Source: Prepared on Erdas Imagine 9.1 and Arc GIS 9.2 using satellite images of 1992 taken form WWF-Lahore

Map 3: Vegetation cover of 2000
Source: Prepared on Erdas Imagine 9.1 and Arc GIS 9.2 using satellite images of 2000 taken form WWF-Lahore

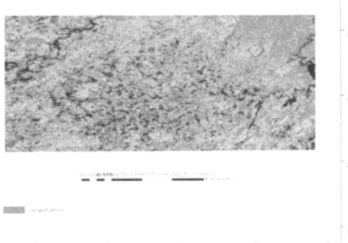

Map 4: Main Kanshi River and its stream network in study area
Source: Prepared on Arc GIS 9.2 from satellite image

Graph 3: Shows the opinion of local community about depletion of water table

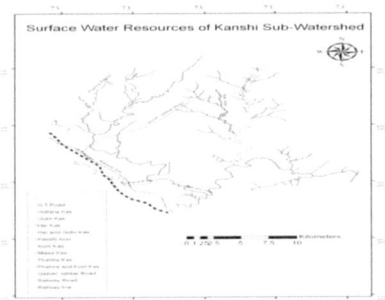

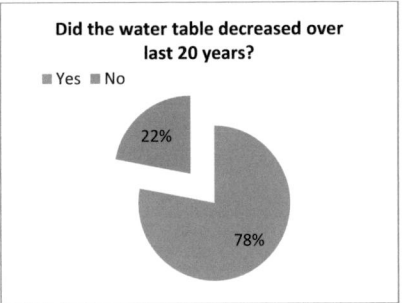

According to the data analysis there are total six local streams namely: Har", "Kurri", "Missa", "Guliana" "Phahna", and "Gulin" which recharge the ground water and contribute water to main Kanshi river. The total discharge of Kanshi river from 1990-2000 was 79.70m³/s and the discharge of second decade (2001-2010) was 44.508 m3/s, the total discharge of 35.193 m3/s is decreased during last 20 years. These changes in water flows also refer to the change in precipitation patterns which could be the impact of temperature variations due to loss of vegetation cover over the last 20 years. The human activities are also the source of ecosystem alteration and destruction. As a result the water table fell from 300 feet to 80 feet in different regions of Kanshi catchment (Gujar Khan).

4. Discussion and Conclusion

During protracted drought, the competition between human and ecological water uses is sharply accentuated. Changes in ground-water use may affect aquifer response more profoundly than climate change associated with modern global warming (Loáiciga 2003)

Annual ground water recharge amounts were found to vary linearly with precipitation. Even in high precipitation years, the infiltration capacity of the watersheds was not exceeded (Nichols and Verry 2001) .Forest clearing and lack of institutional involvement for betterment of environment is causing severe environmental issues like uncertain rainfalls or temporary drought conditions. These conditions have significantly affected the discharge of Kanshi River. Anthropogenic activities are disturbing the natural eco-system and accelerating the hazards expected by climate change. The seasonal streams and rainfall are the major sources of ground water recharge in the study area and due to loss of vegetation cover the precipitation trends are changed and annual flow of these streams is also decreased resulting in lowering of ground water table.

References:

Haslam, S. M. (1995). "Cultural variation in river quality and macrophyte response." Acta Botanica Gallica **142**(6): 595-599.

Loáiciga, H. A. (2003). "Climate Change and Ground Water." Annals of the Association of American Geographers **93**(1): 30-41.

Nichols, D. S. and E. S. Verry (2001). "Stream flow and ground water recharge from small forested watersheds in north central Minnesota." Journal of Hydrology **245**(1–4): 89-103.

Sahin, V. and M. J. Hall (1996). "The effects of afforestation and deforestation on water yields." Journal of Hydrology **178**(1–4): 293-309.

Xin, Z.-b. and X.-x. Yu (2009). "Impact of vegetation restoration on hydrological processes in the middle reaches of the Yellow River, China." Forestry Studies in China **11**(4): 209-218.